Name: _____

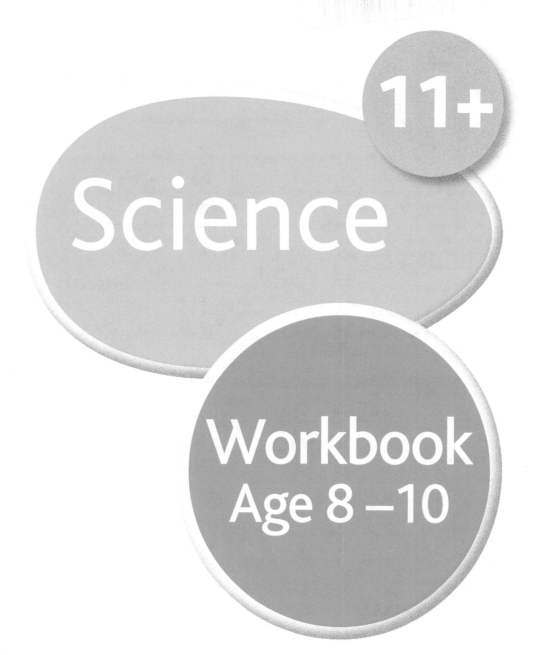

11+

Science

Workbook
Age 8 – 10

Sue Hunter
Jenny Macdonald

GALORE
PARK

AN HACHETTE UK COMPANY

About the authors

Sue Hunter has very recently retired as a full-time science teacher but she continues to be very involved in Science education. She is a member of the Common Entrance setting team and a governor of local primary schools. Sue has written extensively, including Galore Park's KS2 Science textbooks and 11+ Science Revision Guide.

Jenny Macdonald has had a happy and fulfilling career as a teacher, teaching in both state and private schools. She has recently retired having spent the last eighteen years teaching science in a local prep school. In the last few years she has contributed to Galore Park's KS2 Science textbooks.

The Publishers would like to thank the following for permission to reproduce copyright material.

Photo credits
p8 © Danilo Ascione - Fotolia © **p12** isabelle_bonaire - Fotolia.com **p56** © NASA - Yuri Arcurs - Fotolia.com

Acknowledgements
Every effort has been made to trace all copyright holders, but if any have been inadvertently overlooked, the Publishers will be pleased to make the necessary arrangements at the first opportunity.

Although every effort has been made to ensure that website addresses are correct at time of going to press, Galore Park cannot be held responsible for the content of any website mentioned in this book. It is sometimes possible to find a relocated web page by typing in the address of the home page for a website in the URL window of your browser.

Hachette UK's policy is to use papers that are natural, renewable and recyclable products and made from wood grown in well-managed forests and other controlled sources. The logging and manufacturing processes are expected to conform to the environmental regulations of the country of origin.

Orders: Teachers please contact Hachette UK Distribution, Hely Hutchinson Centre, Milton Road, Didcot, Oxfordshire, OX11 7HH. Telephone: (44) 01235 400555. Email: primary@hachette.co.uk. Lines are open from 9 a.m. to 5 p.m., Monday to Friday.

Parents, Tutors please call: 020 3122 6405 (Monday to Friday, 9:30 a.m. to 4.30 p.m.).
Email: parentenquiries@galorepark.co.uk

Visit our website at www.galorepark.co.uk for details of revision guides for Common Entrance, examination papers and Galore Park publications.

ISBN: 978 1 5104 2982 6

First published in 2018 by Galore Park Publishing Limited
Hodder & Stoughton Limited
An Hachette UK Company
Carmelite House
50 Victoria Embankment
London EC4Y 0DZ

www.galorepark.co.uk

Impression number   10 9 8 7 6 5 4 3

Year   2024 2023 2022

Illustrations on p22 and p37 by Oxford Designers & Illustrators.

Illustrations on the following pages by Aptara, Inc.: p7, p9 (top-left), p36, p49 (all), p59 (all)

Illustrations on all other pages by Integra Software Services Ltd.

Typeset in India

Printed in the UK

A catalogue record for this title is available from the British Library.

Contents and progress record

Use these pages to record your progress. Tick the 'Completed' box when you are confident with each skill and record your scores in the 'Score' boxes.

Introduction

This workbook is divided into nine chapters matching Galore Park's Science Year 5 textbook, so you can use it alongside the textbook or on its own.

Each chapter begins with a 'What I already know' section, to remind you about the topic and to provide you with a quick check of your prior learning before you start work. You will then have the opportunity to practise a wide range of scientific skills, including:

- thinking about similarities and differences
- contrasting two life cycles
- plotting a horizontal bar graph
- reading information from a line graph
- looking for patterns in data
- developing your scientific vocabulary
- carrying out research

and so on.

You can easily identify the skill you are going to practise by reading the 'Skill' box at the top of each page.

The 'Skill check' box at the end of each page will encourage you to assess your confidence with each skill.

Each chapter finishes with a quick quiz and a check list to help you assess your progress through the 11+ syllabus and National Curriculum content.

The **contents and progress record** (pages 3–5) helps you keep track of your progress. When you are confident you have mastered each skill, tick the 'Completed' box and record your total score in the 'Score' box.

You might find yourself using this book in the classroom, or for homework, or you might work through it independently at home. It is linked broadly to the 11+ Common Entrance examination syllabus and the National Curriculum, but it will help you to develop key skills and reinforce your learning if you are preparing for any of the following tests:

- Common Entrance 11+ in Science
- 11+ entry tests set by individual schools
- end-of-year tests
- pre-tests.

Answers are provided in a pull-out section in the middle of the book.

1 Life cycles

What I already know

1 Label the plant parts on the diagram. Draw straight lines to join each plant part to its function.

Plant parts   **Function**

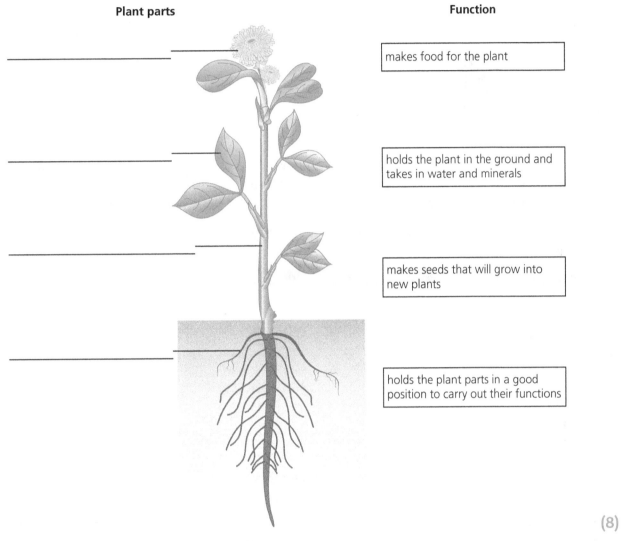

| makes food for the plant |

| holds the plant in the ground and takes in water and minerals |

| makes seeds that will grow into new plants |

| holds the plant parts in a good position to carry out their functions |

(8)

2 Use words from the box to complete the following sentences.

> life cycle   germination   pollination   fertilisation   seed dispersal

(a) Transfer of pollen between flowers by insects or the wind is called _____. (1)

(b) The fusing (joining together) of a pollen grain with an egg cell is called

_____. (1)

(c) Seeds need to travel away from the parent plant. This is _____. (1)

(d) The first stage of growth of a new plant from a seed is called _____. (1)

(e) All these processes together represent the plant's _____. (1)

Round and round

Thinking about similarities and differences

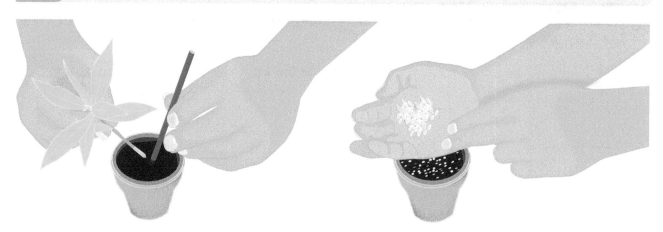

1 Tomato plants can reproduce in two ways. A new plant can be grown from a seed, which is known as sexual reproduction. Alternatively, a gardener can take a small part of a plant, called a cutting, and grow a new plant from it. This is known as asexual reproduction.

Look carefully at the table below. Think about what you know about how plants grow. Place ticks in the columns to show whether each statement refers to a cutting, a seed or both.

	Cutting	Seed
Has a single parent		
Is formed by a pollen grain fertilising an egg cell		
The first things to grow are roots		
Uses food stored in seed		
Needs water		
Needs light		
Needs oxygen		
Needs warmth		
Can carry out photosynthesis immediately		

(9)

2 This cat has just had a kitten. Think about how reproduction in a cat compares with sexual and asexual reproduction in a tomato plant. Complete the following sentence.

Cat reproduction is most similar to growing a tomato plant

from a _____ because _____

_____.

_____ (2)

Skill check

I can identify similarities and differences between sexual and asexual reproduction. ☐

Animal life cycles: insect and frog

Skill   Contrasting two life cycles

1 The life cycles of a Red Admiral butterfly and a common frog are shown below. Compare these two life cycles by filling in the spaces in the table. You may need to do some research to find some of the answers.

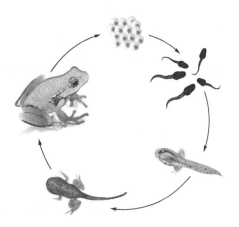

Red Admiral butterfly		Common frog
	Where they lay eggs	
	When they lay their eggs	
	What the young feed upon	
	Name of the young	
	How the adult will be different from the young	
	Approximate time from egg laying to adult form	
	What the adult eats	

(16)

2 Give one way in which these two life cycles are similar.

_____ (1)

Skill check

I can identify the similarities and differences between two life cycles. ☐

Animal life cycles: eggs

Plotting a horizontal bar chart

Looking for patterns in data

There are thousands of different species of birds in the world. Birds come in many different sizes, from the tiny bee humming bird to the ostrich. All birds lay hard-shelled eggs.

The table below shows some data about birds and their eggs.

Bird species	Approximate length of adult, in cm	Approximate length of egg, in cm	Time to hatching, in days
Bee hummingbird	5.5	0.6	22
Kingfisher	18	2.2	20
Emperor penguin	120	12.4	63
Ostrich	230	15	39

1   On the axes below, plot a horizontal bar chart to show the approximate size of each bird's egg.

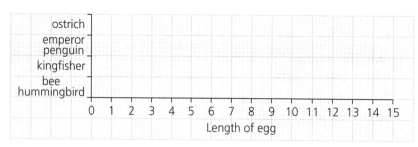

(4)

2   Look carefully at your bar chart and the data in the table about the sizes of each bird species and their eggs. What pattern can you see in the data?

_____

_____ (2)

3   (a)   Now look at the data about the time each egg takes to hatch. One bird's eggs take much longer to hatch than would be expected for their size. Which bird is this?

_____ (1)

   (b)   Suggest a reason why this bird's eggs might take so long to hatch. You may need to do some research to help you.

_____

_____ (2)

Skill check

I can show data in the form of a horizontal bar chart. ⬚

I can identify a simple pattern in data shown in a table or bar chart. ⬚

Life cycles: words and definitions

Skill   Developing vocabulary

In a glossary or dictionary, each word is given a definition. Write the correct word from the box to match each of the definitions below.

asexual	hatching
fertilisation	metamorphosis
fledging	pupa
germination	sexual
imago	tadpole

1 _____ : the first stage of a plant's growth from a seed   (1)

2 _____ : the part of an insect's life cycle where metamorphosis
                     takes place   (1)

3 _____ : when a baby bird leaves the nest for the first time   (1)

4 _____ : the flying adult form of an insect such as a butterfly   (1)

5 _____ : (reproduction) involving just one parent, for example
                     taking a cutting   (1)

6 _____ : when a baby animal emerges form an egg   (1)

7 _____ : a complete change of body form during a life cycle,
                     for example tadpole to frog   (1)

8 _____ : the joining together of a sperm cell or pollen grain and an
                     egg cell   (1)

9 _____ : the young form of an amphibian   (1)

10 _____ : (reproduction) involving two parents   (1)

Skill check

I know and understand the vocabulary for this topic. ☐

Research project: sea turtle life cycle

Use books or the internet to find out about the life cycle of a sea turtle. Try to find as many interesting facts as possible. You can present your work in whatever way you choose.

_____

_____

_____

_____

_____

_____

_____

_____

_____

_____

_____

_____

_____

Bibliography
Keep a note of the book titles and websites that you used for your research in the space below.

_____

_____

Skill check

I can find appropriate sources of information for my research. ☐

I have kept a record of my sources in the bibliography. ☐

Quick quiz: life cycles

Write *short* answers to the following questions to check how much you remember about life cycles. You do not need to write full sentences.

1 What word is used to describe reproduction involving two parents?

_____ (1)

2 What word is used to describe reproduction involving only one parent?

_____ (1)

3 What is meant by the term 'fertilisation'?

_____

_____ (1)

4 What name is given to the complete change of form experienced by many amphibians and insects during their life cycle?

_____ (1)

5 What name is given to the eggs of frogs?

_____ (1)

6 Describe briefly the changes that occur in a frog as it develops from a tadpole to an adult frog.

_____

_____ (3)

7 What name is given to the part of the life cycle of a butterfly where metamorphosis occurs?

_____ (1)

8 What is meant by the term 'imago'?

_____ (1)

9 Fill in the gaps in the following passage:

Birds lay hard-shelled eggs in a _____. The parent birds will

_____ the eggs until the baby birds _____. The

parents will feed the babies in the nest until they _____ (leave the nest).

(4)

Turn over to the next page for the final question in this quiz.

10 Underline the correct words from the brackets to complete each of the following sentences:

(a) The first stage in the growth of a new plant from a seed is called (fertilisation/germination). (1)

(b) Many trees have small green flowers, showing that they are pollinated by (the wind/insects). (1)

(c) Plants that rely on the wind to disperse their seeds often have fruits that (have wings/are juicy). (1)

(d) New plants can sometimes be made asexually by using (cuttings/seeds). (1)

(Total 18)

Check list: life cycles			
I know	☺	😐	☹
the difference between sexual and asexual reproduction			
that fertilisation is the fusing (joining together) of male and female sex cells in sexual reproduction			
about the life cycle of an amphibian, e.g. a frog			
about the life cycle of an insect, e.g. a butterfly			
about the life cycle of a bird, e.g. a robin			
about how plants reproduce sexually and asexually			
I can	☺	😐	☹
describe differences between the life cycles of a mammal, an amphibian, an insect and a bird			

2 More about life cycles

What I already know

Here is a picture of a newborn human baby. Think about what this baby will need so that it is happy and healthy. Write your ideas in the boxes. You may add more boxes if you like.

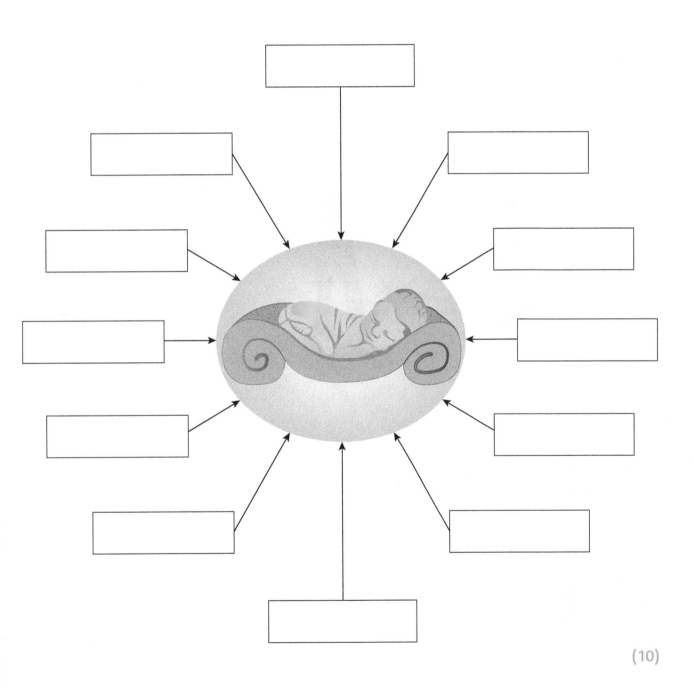

(10)

The human cycle: how old?

The image below shows some stages in the life cycle of humans.

1 Using the words from the box below, label the different stages of the life cycle neatly
   on the image.   (5)

   | adolescent | adult | baby | child | fetus |

2 Next to each of the pictures suggest an approximate age for each of the people shown.   (5)

Skill check

I can use scientific vocabulary to name the stages in the human life cycle.

I have used prior knowledge to suggest the ages shown on a life cycle diagram.

The human cycle: growing baby

| Skill | Reading information from a line graph |

The graph below shows the changes in length of a human baby girl.

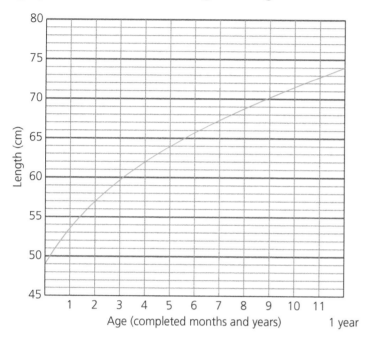

Use the graph to find the answers to the following questions.

1 How long was the baby when she was born? _____ cm.   (1)

2 How long was she at the age of 5 months? _____ cm.   (1)

3 How long was she at 1 year old? _____ cm.   (1)

4 How much did she grow in her first year? Show your working.

_____ cm. (2)

5 (a) Is the growth line steeper between birth and 3 months or between 9 months and 1 year?

_____ (1)

   (b) What does this tell us about how quickly the baby grew during these periods?

_____ (1)

Skill check

I can read information accurately from a line graph. ◯

Other animals: animal gestation periods

Skill	Interpreting a bar chart

The bar chart below shows the gestation periods of some animals.

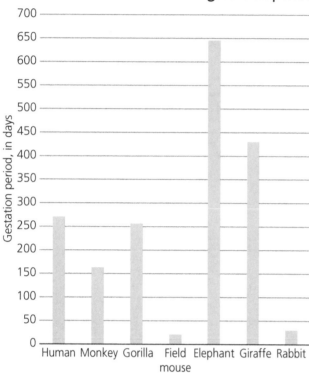

Use the bar chart to help you to answer the following questions.

1 Which of the animals has the longest gestation period?

_____ (1)

2 Which of the animals has the shortest gestation period?

_____ (1)

3 Which of the animals have a longer gestation period than the human?

_____ (2)

4 Circle words in the brackets to complete the following sentence.
  The (bigger/smaller) the animal, the (shorter/longer) the gestation period. (2)

5 Use the bar chart to estimate the gestation period of a monkey.

_____ (1)

6 Use information from the bar chart to predict the gestation period of a chimpanzee.

_____ (1)

Skill check

I can read and understand information from a bar chart. ☐

Human life cycle: wordsearch

Skill Developing vocabulary

Find words in the wordsearch to complete the definitions below. Highlight each word neatly when you find it and make sure that you check the spelling when writing it.

F	G	U	U	T	E	R	U	S	G	I	F	U
A	N	Z	Y	L	L	I	O	V	F	R	T	M
W	W	X	A	H	M	M	F	N	A	K	T	B
W	S	V	D	L	E	A	O	B	L	D	Y	I
J	T	W	O	E	N	M	M	L	L	L	W	L
O	F	T	L	O	S	M	P	G	O	E	L	I
V	U	E	E	W	T	A	U	N	P	Q	U	C
A	C	S	S	N	R	L	Y	Y	I	N	J	A
R	R	T	C	A	U	S	X	T	A	D	Q	L
I	F	E	E	W	A	F	X	Q	N	V	T	P
E	X	S	N	R	L	M	U	Z	B	L	G	B
S	K	W	C	S	V	D	L	N	R	P	I	Y
D	I	M	E	J	W	P	E	N	I	S	F	A
L	V	G	E	S	T	A	T	I	O	N	Y	B

1 During _____, changes occur in the body as a child becomes an adult.

2 During the _____ cycle, the lining of the uterus is shed and renewed.

3 The _____ is the organ in a woman's body where a fetus develops.

4 The _____ tube leads from an ovary to the uterus.

5 The places in a woman's body where eggs are made are called the _____.

6 The _____ are the parts of a man's body that make sperm cells.

7 The structure joining the fetus to the placenta in the uterus is the _____ cord.

8 The _____ period is the length of time a fetus spends developing in the uterus.

9 The vertebrate animals that feed their young on milk are _____.

10 The part of a man's body through which sperm and urine leave the body is the _____.

(10)

Skill check

I know and understand the vocabulary for this topic. ☐

Quick quiz: more about life cycles

Write *short* answers to the following questions to check how much you remember about life cycles. You do not need to write full sentences.

1 Rewrite the following stages in the human life cycle in the correct order: adolescent, adult, baby, child, fetus.

_____ (1)

2 Underline the correct option to complete the following sentence.
The gestation time of a human baby is about

| 6 months | 9 months | 12 months | 24 months |

(1)

3 Underline the correct option to complete the following sentences.
(a) The mass of a newborn human infant is closest to

| 300 g | 3 kg | 30 kg | 300 kg |

(1)

(b) The mass of a Year 5 pupil is closest to

| 300 g | 3 kg | 30 kg | 300 kg |

(1)

4 Suggest one feature that is common to most animal life cycles.

_____ (1)

5 Name two changes that are experienced by both boys and girls during adolescence.

_____ (2)

6 Name two changes that only girls experience in their bodies during adolescence.

_____ (2)

7 Name two changes that only boys experience in their bodies during adolescence.

_____ (2)

(Total 11)

Check list: more about life cycles

I know	☺	😐	☹
the stages of the human life cycle			
that the gestation period of a human fetus is nine months			
about how the length and mass of a human infant changes as it grows			
some features of life cycles that are common to all animals			
that during adolescence, physical and emotional changes take place			
that during adolescence, both girls and boys develop hair around the sex organs and experience mood changes			
that changes in adolescence for girls include the development of breasts and the menstrual cycle			
that changes in adolescence for boys include facial hair growth, deepening voice and increase in the size of penis and testes			
I can	☺	😐	☹
make comparisons between the life cycles of different mammals, including gestation periods			

The picture shows a pond habitat.

1 Name three things that a habitat must provide for an animal to survive there successfully.

_____

_____ (3)

2 Label the picture to show at least one example of each of the following life processes: reproduction, nutrition (feeding), movement and growth. (4)

3 Label the picture to show at least two ways in which human activity has had an impact on this habitat. (2)

What is a habitat?: mole home for sale

| Skill | Demonstrating understanding through creative writing |

Read the following information about moles carefully.

Moles live almost entirely underground in a network of tunnels. They dig the tunnels with their strong front feet. Every now and again they dig a vertical shaft and push up the spare soil, making a mole hill. This makes them unpopular with gardeners and farmers. Moles prefer eating earthworms but will also eat beetles, fly larvae, and slugs. They may store earthworms with their heads bitten off near their nests, which are lined with grass. If one mole dies or moves away, another mole will move into the territory.

Imagine you are a mole who has dug a wonderful tunnel system. You have decided to move. Write a description of your home to persuade another mole to take it over. Think about what might make an area attractive to a mole (for example, soil type, food supply, land use above the tunnels, etc.). Make your description really persuasive.

_____

_____

_____

_____

_____

_____

_____

_____

_____

_____

_____

_____

Skill check

I have written an imaginative passage showing understanding of information from a non-fiction text. ☐

Adaptation and habitats: anagrams

The anagrams below are words that relate to the topic of adaptation and habitats.

Unscramble each word and write it in the correct place next to its definition in the table below.

1 hirabtene
2 tatibah
3 tamiegr
4 norlnctua
5 evintomnren
6 tadpatnoia
7 resthel
8 cofauglmae

Definition	Word		Anagram number
An area where an animal or plant lives and which provides food, water and a place to breed			
The surroundings, the whole of the natural world			
To move from one place to another during the year, usually for food or to find suitable breeding grounds			
Active at night			
A place that provides safety and protection from danger or bad weather			
To slow body processes to a complete minimum and sleep when the weather is very cold			
Feature of the body of an animal or plant that improves its ability to survive in its habitat			
Feature of the body (shape and/or colour) that allows it to blend into its background			

(8)

Skill check

I know and understand the vocabulary for this topic. ☐

Adaptation: capybara

Skill   Reading text for information

The first European naturalists to visit South America named capybaras 'water pigs'. However, they are neither pigs, nor are they totally aquatic. They are, however, the largest of all the rodents, with an average mass of 50 kg.

They have a large barrel-shaped body and move rather slowly and clumsily. The body is covered in long, bristly hairs, usually reddish brown in colour. They have no tail and their front legs are shorter than their back legs. Their slightly webbed toes, four on the front feet and three on the back, make them very strong swimmers. The nostrils, eyes and ears are on the top of the head, and can remain above the surface when the animal is swimming.

Capybaras are herbivores and feed mainly on grasses found in or near water. They have two pairs of large incisors and can eat very short grasses, which they grind with their molar teeth.

Capybaras live in groups of between 10 and 100. When a predator is seen, they give an alarm bark, and the reaction of the group is to rush into the water.

1 Explain how the feet of the capybara have adapted to life in water.

_____ (1)

2 Describe the features of the head of a capybara that make it well-adapted to life in water.

_____

_____ (3)

3 Which word in the text tells us that capybaras eat plants? _____ (1)

4 Two types of tooth are mentioned in the passage. Which type is used for:

   (a) biting the grasses _____ (1)   (b) chewing the grass _____ (1)

5 Suggest why early naturalists named capybaras 'water pigs'.

_____

_____ (3)

6 Describe in your own words why a capybara might choose to run into water rather than running away from a predator across land.

_____

_____ (2)

Skill check

I can use information from text to answer questions.   ☐

Research project: swallows

British swallows fly to South Africa during the cold British winter and return to Britain for the warmer summer months. This is called 'migration'.

Use books and the internet to find out about swallow migration, and use your research to complete the factsheet below.

Swallow migration factsheet

Distance travelled in one day:

_____

How fast they travel:

_____

Approximate total distance travelled:

_____

Main time of travel (circle your answer): day/night

Approximate date leaving Britain:

_____

Approximate return date:

_____

On the map on the right mark the routes of the migrating swallows between Britain and South Africa

Name some countries that the swallows fly across.

_____

_____

How do the swallows prepare for their journey?

_____

_____

Describe some of the dangers the birds may face on their long journey.

_____

_____

Skill check

I can use books and the internet appropriately to find out about a topic. ☐

Quick quiz: adaptation and habitats

Write *short* answers to the following questions to check how much you remember about adaptation and habitats. You do not need to write full sentences.

1 (a) What is meant by the term 'habitat'?

_____ (1)

   (b) What must a habitat provide for the organisms that live there?

_____

_____ (3)

2 (a) What word is used to describe the features of an animal or plant that help it to survive in its habitat?

_____ (1)

   (b) Describe one such feature of an animal or plant that you have studied and say how it helps the organism to survive.

_____

_____ (2)

3 (a) What is meant by the term 'nocturnal'?

_____ (1)

   (b) Give an example of a nocturnal animal.

_____ (1)

4 Give two reasons why an animal might migrate.

_____ (2)

5 Hedgehogs are seldom seen during the winter. Why is this?

_____ (1)

6 Briefly describe a habitat that you have studied.

_____

_____

_____ (3)

Turn over to the next page for the final question in this quiz.

7 (a) Name one animal that is found in the habitat you have studied.

_____ (1)

(b) Describe the features of this animal as follows:

Size: _____

Shape and colour: _____

How it moves: _____

How it feeds: _____ (4)

(Total 20)

Check list: adaptation and habitats			
I know	☺	😐	☹
that a habitat is a place that provides food, water and a place to breed for the organisms living there			
that an animal or plant will have specific features that improve its ability to survive in its habitat, and that these are called adaptations			
that some animals and plants have specific features, such as colouration or body shape, that allow them to blend into the background, and that this is called camouflage			
that some animals are nocturnal (active by night)			
that animals have a variety of responses to changes in the seasons, including migration and hibernation			
I can	☺	😐	☹
describe the main features of a habitat that I have studied			
describe the features of plants and animals in a habitat that I have studied			

4 Properties of materials

What I already know

When we make things, we choose materials to suit the job. To do this we consider the properties of the materials.

The table shows some materials and things that have been made from them.

Three properties of materials are represented in the Venn diagram.

Think carefully about each material and the thing it was used to make, then write its name into the correct section of the Venn diagram.

Material	is used to make
aluminium	cooking foil
ceramic (china)	tea mug
chalk	blackboard chalk
cotton cloth	shirt
glass	window
paper	this page
iron	railings
waxed cotton	raincoat
plastic	food wrap

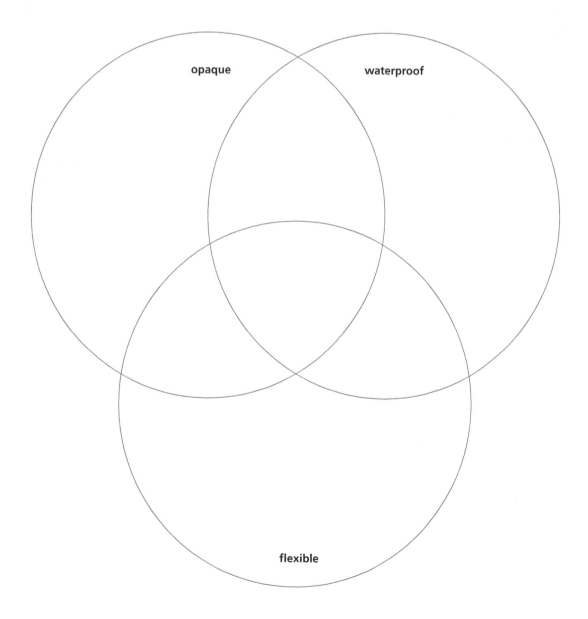

opaque   waterproof

flexible

(9)

Properties of materials

Draw one or two straight lines from each of the items pictured below to link them to properties that are essential for their function (job).

flexible

rigid

strong

hard

soft

absorbent

waterproof

flammable

non-flammable

shiny

thermal insulator

(10)

Skill check

I can select properties that are essential to the function of an item. ☐

Runny races

Skill   Designing a fair test

Using prior knowledge to predict results

Some children are investigating how liquids flow.

They use the apparatus below.

They put a spoonful of a liquid on the top line and measure how long it takes for the liquid to reach the bottom line.

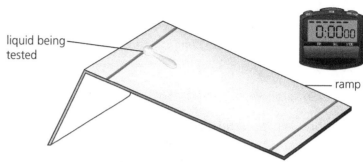

liquid being tested

ramp

Their teacher asks them to design an experiment to compare how fast four different liquids run down the slope.

The liquids are: water, tomato ketchup, golden syrup and oil.

1   Suggest how the children could carry out a fair test to compare how fast these four liquids run down the slope. You may add to the diagram above if you wish.

_____

_____

_____

_____

_____

_____

_____ (5)

2   Predict which liquid would flow down the slope slowest. Explain your answer.

_____

_____ (2)

3   Vocabulary challenge
Find a word, beginning with the letter v, that describes the property of 'stickiness' in liquids.

_____ (1)

Skill check

I can design a simple fair test, including identifying control variables. ☐

I can use my existing knowledge and understanding to predict and explain a result. ☐

Testing materials: school bags

Some scientists have been asked to test some fabrics for making school bags. The bags must not wear out too quickly.

The scientists use this apparatus.

The fabric is placed on the base of the apparatus and the block is rubbed back and forth across the fabric until it shows signs of wear. They count the number of rubs.

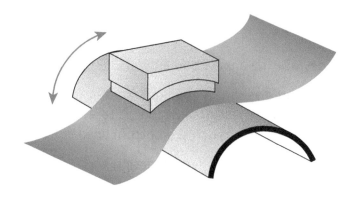

	Fabric	How many rubs (thousands)
A	Canvas with waterproof backing	25
B	Thick leather	40
C	Cotton with waterproof backing	5
D	Nylon	

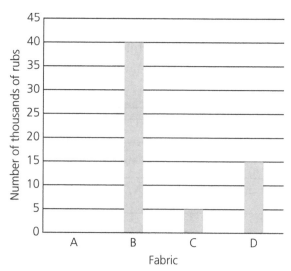

1 Use the bar chart to find and complete the missing value in the table. (1)
2 Use the data from the table to complete the bar chart by drawing in the missing bar for fabric A. (1)
3 How could the scientists check that their results are reliable?

_____ (1)

4 Fabric A is chosen for the bags. Suggest why the school thought that this was the best choice.

_____

_____ (2)

Skill check

I can read and complete a bar chart. ☐

I can use data from a table or bar chart to suggest conclusions. ☐

Answers

1 Life cycles

What I already know (page 7)

1. (labels from top to bottom:) flower – makes seeds that will grow into new plants
   leaf – makes food for the plant
   stem – holds the plant parts in a good position to carry out their functions
   root – holds the plant in the ground and takes in water and minerals (8)
2. (a) pollination (1)
   (b) fertilisation (1)
   (c) seed dispersal (1)
   (d) germination (1)
   (e) life cycle (1)

Round and round (page 8)

1.

	Cutting	Seed
Has a single parent	✓	
Is formed by a pollen grain fertilising an egg cell		✓
The first things to grow are roots	✓	✓
Uses food stored in seed		✓
Needs water	✓	✓
Needs light	✓	
Needs oxygen	✓	✓
Needs warmth	✓	✓
Can carry out photosynthesis immediately	✓	

(9)

2. Cat reproduction is most similar to growing a tomato plant from a **seed** because **both involve sexual reproduction/two parents**. (2)

Animal life cycles: insect and frog (page 9)

1.

Red Admiral butterfly		Common frog
On (nettle) leaves	Where they lay eggs	In water/pond
Spring	When they lay their eggs	Spring
(Nettle) leaves	What the young feed upon	Algae and then small invertebrates
Caterpillar	Name of the young	Tadpole

Red Admiral butterfly		Common frog
Adult has wings, an exoskeleton/ hard outer skin, three body parts (Accept at least two valid suggestions.)	How the adult will be different from the young	Adult has legs, no tail, lungs, long tongue for feeding (Accept at least two valid suggestions.)
About six weeks	Approximate time from egg laying to adult form	About 13–15 weeks
Nectar	What the adult eats	Flies and other insects

(16)

2. Both life cycles involve metamorphosis. (1)
   (Accept valid alternatives.)

Animal life cycles: eggs (page 10)

1. One mark for each correctly drawn bar (4)

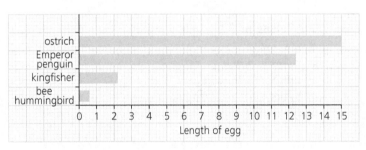
Length of egg

2. The larger/smaller the bird, the larger/smaller its eggs. (2)
3. (a) Emperor penguin (1)
   (b) Suggestions could include: It is very cold in the penguin breeding area so the chick develops more slowly./The chick needs to hatch out when the weather starts to warm up to give it a better chance of survival. (2)
   (Allow sensible, well-reasoned alternatives.)

Life cycles: words and definitions (page 11)

1. germination (1)
2. pupa (1)
3. fledging (1)
4. imago (1)
5. asexual (1)
6. hatching (1)
7. metamorphosis (1)
8. fertilisation (1)
9. tadpole (1)
10. sexual (1)

Research project: sea turtle life cycle (page 12)

Pupils should carry out research to find out information relating to the life cycle of a sea turtle of any species, working at a level appropriate to age and ability. They should focus on reproduction and not include other information, however interesting! They could present their work as notes, a poster, a PowerPoint presentation or as a talk to the class. They should record their sources in the bibliography.

Quick quiz: life cycles (page 13)

1 sexual   (1)
2 asexual   (1)
3 fusing together of a male sex cell (pollen or sperm) and an egg cell   (1)
4 metamorphosis   (1)
5 (frog) spawn   (1)
6 loses its tail, grows legs (back then front), gills change to lungs   (3)
7 pupa
   (accept chrysalis)   (1)
8 flying adult form of an insect   (1)
9 nest, sit on/incubate, hatch, fledge   (4)
10 (a) germination   (1)
   (b) wind   (1)
   (c) have wings   (1)
   (d) cuttings   (1)
                                                        (Total 18)

2 More about life cycles

What I already know (page 15)

Suggestions might include:

- food/milk
- warmth/clothing
- safe home/environment
- hygiene/bathing/nappy change
- stimulation/entertainment/toys
- love/care
- parents/carer/mother and father
                                                             (10)

The human cycle: how old? (page 16)

1 Stages correctly identified   (5)
2 Approximate ages might be:
   fetus – minus 6 months/15 weeks gestation. (Allow 0)
   baby – between 6 months and 1 year
   child – 5–10 years
   adolescent – 13–16 years
   adult – 20+ years   (5)

The human cycle: growing baby (page 17)

1 49 cm   (1)
2 64 cm   (1)
3 74 cm   (1)
4 25 cm (2) (one for working, one for correct answer)
5 (a) steeper during the first three months than between nine months and one year   (1)
   (b) The baby grew more rapidly in the first three months than in the last three months of the year.   (1)

Other animals: animal gestation periods (page 18)

1 elephant   (1)
2 field mouse   (1)
3 elephant and giraffe   (2)
4 The bigger the animal, the longer the gestation period. **or** The smaller the animal, the shorter the gestation period.   (2)
5 160   (1)
   (Accept answers in the range 155–170)
6 240 days   (1)
   (Accept any answer greater than 164 [monkey] and less than 270 [human].)

Human life cycle: wordsearch (page 19)

1 adolescence
2 menstrual
3 uterus
4 fallopian
5 ovaries
6 testes
7 umbilical
8 gestation
9 mammals
10 penis   (10)

Quick quiz: more about life cycles (page 20)

1 fetus, baby, child, adolescent, adult   (1)
2 nine months   (1)
3 (a) 3 kg   (1)
   (b) 30 kg   (1)
4 two parents (one male and one female)   (1)
5 hair around sex organs, mood swings, spots   (2)
   (Accept any two valid answers.)
6 start menstruating (periods), breasts develop, hips widen (2)
   (Accept any two valid answers.)
7 voice deepens, facial hair, penis and testes increase in size   (2)
   (Accept any two valid answers.)   (Total 11)

3 Adaptation and habitats

What I already know (page 22)

1 supply of food and water, safe space to shelter/hide from predators /rear young, oxygen to breathe   (3)
   (Allow any three valid suggestions.)
2 At least one example of each of the four life processes correctly identified and labelled, e.g:
   reproduction: birds in nest, toads, caterpillars and butterflies
   growth: plants growing, baby and adult toads, baby and adult birds
   nutrition: birds feeding young, caterpillar eating leaf
   movement: bird/butterfly flying, toad jumping   (4)
3 trolley dumped in the water, pollution from factory/cars, car park covering land   (2)
   (Accept any two valid suggestions correctly labelled.)

What is a habitat?: mole home for sale (page 23)

The pupil should write a persuasive description of a mole tunnel network showing understanding of conditions that would be favourable to the mole (e.g. loamy soil containing many earthworms, beetle larvae, etc. and easy to dig, well-drained and not likely to flood, grass above to line the nest, not in a garden or field where the owner might take exception to mole hills).

Adaptation and habitats: anagrams (page 24)

Definition	Word	Anagram number
An area where an animal or plant lives and which provides food, water and a place to breed	habitat	2
The surroundings, the whole of the natural world	environment	5
To move from one place to another during the year, usually for food or to find suitable breeding grounds	migrate	3
Active at night	nocturnal	4
A place that provides safety and protection from danger or bad weather	shelter	7
To slow body processes to a complete minimum and sleep when the weather is very cold	hibernate	1
Feature of the body of an animal or plant that improves its ability to survive in its habitat	adaptation	6
Features of the body (shape and/or colour) that allow it to blend into its background	camouflage	8

(8)

Adaptation: capybara (page 25)

1 webbed toes (1)
2 eyes, nostril and ears are on the top of the head so that they remain above the surface when swimming. (3)
3 herbivore (1)
4 (a) incisors (1)
  (b) molars (1)
5 They have barrel-shaped bodies and short legs like a pig and they live largely in water. (3)
6 They cannot run fast because they have heavy bodies and short legs, and they can swim well. (2)

Research project: swallows (page 26)

Distance travelled in one day: 320 km/200 miles
How fast they travel: 27–34 km per hour/17–21 miles per hour
Approximate total distance travelled: 12 800 km/8000 miles each way, depending on route taken
Main time of travel: night
Approximate date leaving Britain: September/October
Approximate return date: April/May
Map: main routes are via France, Spain, Morocco, then across the Sahara or down the west coast of Africa, avoiding the Sahara.
Countries: France, Spain, Morocco, the African countries (e.g. Niger, Nigeria, Cameroon, Zaire, Angola, etc.)
Preparation: they fatten up by eating plenty of insects before leaving and then refuel as necessary before each stage of the journey.
Dangers: weather, starvation, exhaustion, humans with guns or nets, predatory birds

Quick quiz: adaptation and habitats (page 27)

1 (a) a place where an animal or plant lives (1)
  (b) food, water and a place to breed/shelter (3)
2 (a) adaptations (1)
  (b) Answer should describe at least one feature of the chosen animal/plant and say how it helps it to survive. (2)
3 (a) active at night (1)
  (b) e.g. hedgehog, mole, bat, etc. (1)
4 to find food or to a suitable breeding area or warmer climate (2)
5 they hibernate (1)
6 a brief description of the studied habitat (e.g. a small pond with shallow beach area and surrounded by vegetation/a deciduous woodland containing mainly beech trees with some shrubby undergrowth) (3)
7 (a) a suitable animal from the described habitat (1)
  (b) accurate description of the named features (4)

(Total 20)

4 Properties of materials

What I already know (page 29)

(9)

Properties of materials (page 30)

- raincoat: flexible/waterproof
- firewood: flammable
- swimming towels: absorbent/flexible/soft
- mirror: shiny/rigid
- arm chair: soft/strong/non-flammable
- bridge: strong/rigid
- tea cosy: thermal insulator/flexible
- fireman's jacket: non-flammable/flexible/strong
- bucket: waterproof/rigid
- hammer: strong/rigid/hard

(10)

(One or two links for each item. The links should only be with properties that are essential for the function of the item.)

Runny races (page 31)

1   either:
    Place a spoonful of a liquid on the top line. Time how long it takes to reach the bottom line. Repeat with the other liquids in turn making sure that the volume of liquid, start position and end position are the same.
    or
    Put the board down flat. Put equal volumes of each liquid on the start line side by side. Lift the board to start them all at the same time and record the order in which they reach the bottom line, making sure that the same end-point is used for each.   (5)
2   Accept either tomato ketchup or golden syrup. These liquids are thick/sticky/less runny so will flow more slowly.   (2)
3   viscous/viscosity   (1)

Testing materials: school bags (page 32)

1   15 thousand rubs   (1)
2   bar showing 25 thousand rubs   (1)
3   repeat test three times to check repeatability of results   (1)
4   Accept any valid suggestions (e.g. A is waterproof/more flexible than the leather, more possibility for colour choice/cost, etc.)   (2)

Quick quiz: properties of materials (page 33)

1   what an object is made of   (1)
2   It is transparent/translucent, rigid, waterproof/windproof, strong.   (3)
3   (a) e.g. strong/hard/easily available   (1)
    (b) e.g. waterproof/strong/rigid/thermal insulator/inexpensive   (1)
    (c) e.g. flexible/waterproof/strong   (1)
    (different choice for each item)
4   metal and non-metal   (2)
5   magnetic   (1)
6   iron   (1)
7   because steel contains iron   (1)
8   non-magnetic   (1)
9   Hang each bag up.
    Place masses inside one until it tears.
    Repeat with the other bag using the same masses.
    Compare the results.
    The one that hold the most is the strongest.   (5)

(Total 18)

5 Reversible changes

What I already know (page 35)

1   chocolate: solid to liquid – melting
2   water: liquid to solid – freezing
3   water: gas to liquid – condensation
4   water: liquid to gas – evaporation   (12)

Changes of state: the water cycle (page 36)

C   The heat from the Sun warms the **water** in the sea. Some of the water **evaporates** and rises up into the sky as water vapour.
B   As the water **vapour** rises, it cools and **condenses** to form tiny droplets of liquid water. These come together to form clouds.
D   Winds blow the clouds over the land. When they meet mountains, they rise and **cool** even more. This makes the water droplets bigger and they fall as rain, sleet or snow. This is called **precipitation**.
A   Most of the water flows into **rivers,** which carry it back to the sea. Some water soaks into the soil to become ground water.
E   Some water may get trapped in crevices in the rocks on the mountain. In winter this water may **freeze**. Water **expands** when it freezes and this will push the crevices apart, **cracking** the rocks into smaller pieces.

(15)

(10 for filling in the gaps correctly plus 5 for matching statements and letters)

Garden hose disaster (page 37)

The pupil should write a clearly worded letter, including a description of the event and an explanation of why it happened. The explanation should include understanding of the facts that water must have been left in the hose and frozen during the winter, that water expands on freezing and that this would be responsible for splitting the hosepipe.   (10)

(10 marks awarded at marker's discretion, with attention to the application of literacy skills as well as science)

Dissolving experiment (page 38)

1

	Time taken to dissolve, in minutes and seconds		
	Test 1	Test 2	Test 3
Group A	0 mins 35 secs	0 mins 26 secs	0 mins 58 secs
Group B	1 min 45 secs	1 min 42 secs	1 min 47 secs

(4)

2   (a) group A   (1)
    (b) make sure that the same mass of salt and same volume of water are used, water at same temperature.   (2)
    (Accept any two reasonable suggestions.)

3   wear safety goggles, take care when handling hot liquids   (1)
    (Accept any one valid suggestion.)
4   100 cm³ of water weighs 100 g.
    100 g water + 10 g of salt = 110 g   (2)

Solutions: crossword (page 39)

Across

1   solution
3   saturated
4   soluble
7   suspension
8   solvent

Down

2   insoluble
5   dissolved
6   solute   (8)

Quick quiz: reversible changes (page 40)

1   The ice will melt back to water when warmed (and can be refrozen).   (1)
2   (a) melting   (1)
    (b) condensing/condensation   (1)
3   The water evaporates (to form water vapour in the air).   (1)
4   boiling   (1)
5   0 °C   (1)
6   It expands when it freezes   (1)
7   a solid unable to dissolve in water/liquid   (1)
8   a (salt) solution   (1)
9   (a) solute   (1)
    (b) solvent   (1)
10 Small crystals will dissolve quicker.   (1)
11 The higher the temperature the faster the solute will dissolve.   (2)
12 Stirring increases the rate of dissolving.   (1)
13 120 g (100 + 20)   (1)
                                                      (Total 16)

6 Separating mixtures

What I already know (page 42)

1   The particles in a soil sample can be separated using sieves.   (1)
2   Iron is a magnetic metal.   (1)
3   A mixture of steel and plastic paper clips could be separated using a magnet.   (1)
4   An example of a mixture is muesli.   (1)
5   Air is a mixture of gases.   (1)
6   When water is heated it evaporates.   (1)
7   A solution contains a liquid and a dissolved solid.   (1)
8   Leaves and sticks will float on water but stones will sink.   (1)

Mixtures are all around us (page 43)

1

Mixtures	Pure materials
air	iron
chocolate	oxygen gas
coffee (drink)	salt (assuming that the salt is pure!)
muesli	
orange squash	
sea water	
soil	

                                                          (10)

2

Mixture	Odd one out	Why
Flour, sand, eggs, butter	e.g. sand	inedible/not part of a cake mix
Water, salt, seaweed, sugar	e.g. seaweed	living
	sugar	not found in sea water
	water	liquid
Gravel, leaves, silt, glass	e.g. leaves	living/once living/float
	glass	man-made/not usually found in soil

                        (3) (Accept valid alternatives.)

Separating mixtures (page 44)

The pupils should wear eye protection. They should pour the salt solution into an evaporating basin. They should place the tripod stand (with the gauze mat on top) and Bunsen burner on a heatproof mat.

Method 1: Place the beaker of water on the tripod stand and balance the evaporating basin on top, light the Bunsen burner safely and heat strongly so that the water in the beaker boils and warms the salt solution.
OR
Method 2: Place the evaporating basin on top of the tripod stand, light the Bunsen burner safely and heat gently. When nearly all the water has evaporated from the solution they should stop heating, to avoid hot crystals spitting out, and allow the crystals to dry and cool.   (10)

Drawing science diagrams (page 45)

1

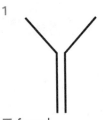

■ funnel

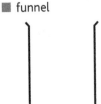

■ beaker

■ conical flask (6)

(Allow one mark for correct diagram format and one for neatness in each.)

2

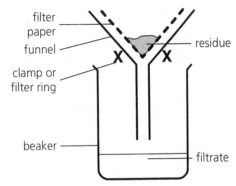

filter paper
funnel
residue
clamp or filter ring
beaker
filtrate

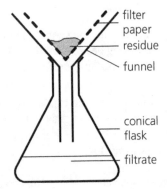

filter paper
residue
funnel
conical flask
filtrate

(5)

(Either diagram is acceptable. Allow two marks for apparatus, two for labelling and one for neatness.)

Quick quiz: separating mixtures (page 46)

1 (a) filtering/filtration (1)
   (b) decanting (1)
2 (a) residue (1)
   (b) filtrate (1)

3 (a) evaporating (1)
   (b) the water evaporates leaving the salt crystals behind (2)
4 to prevent hot salt crystals from spitting out (or burning the salt crystals) (1)

(Total 8)

7 Chemical changes

What I already know (page 47)

- firework exploding – non-reversible (1)
- snowman melting – reversible (although a discussion could be had about whether it would be possible to reform the snow crystals or snowman) (1)
- egg frying – non-reversible (1)
- hair drying – reversible (1)
- sugar dissolving – reversible (1)
- leaves rotting – non-reversible (1)
- wood burning – non-reversible (1)
- trees growing – non-reversible (1)

An important chemical change: combustion (page 48)

1 (a) balance/scales (1)
   (b) measuring cylinder (1)
   (c) thermometer (1)
2 wear safety glasses/goggles (1)
   (Allow valid alternatives.)
3 The measurement labels on the vertical axis are missing. (1)
4

	True	False	Cannot tell
Fuel A heated the water most	✓		
The temperature rise in the water for Fuel B was 20 °C			✓
Fuel C heated the water less than Fuel B		✓	
All the fuels heated the water by more than 10 °C			✓

(4)

Another chemical change: rusting (page 49)

1 Answers here will vary depending on the pupil's thoughts about each method.
   Examples:
   Child 1: This might work if enough sticky tape was used because the tape would keep air and water away from the nail. However sticky tape becomes soggy and would come away from the nail if it gets too wet and then the nail would rust.
   Child 2: This would work if the glue is not water-soluble. PVA glue would wash off and then water and air could get to the nail and make it rusty.

Child 3: If the oil makes a good layer all over the nail it would stop water and air getting to the nail. If it all runs off the nail would not be protected.

Child 4: This method should work very well as the nail varnish would seal the nail from water and air. She would need to make sure that the whole nail was covered. (8)

2 control (1)

Quick quiz: chemical changes (page 50)

1 They form new materials. (1)
2 (a) man-made (synthetic) (1)
  (b) man-made (synthetic) (1)
  (c) naturally occurring (1)
  (d) naturally occurring (1)
3 water and air/oxygen (2)
4 oiling/painting/galvanising/covering with plastic (3)
  (any three needed)
5 a substance that is burned to release useful energy (2)
6 a (non-renewable) fuel formed millions of years ago (from the remains of plants and animals) (2)
7 (a) coal/peat (1)
  (b) oil (1)
  (c) natural gas (1)
8 combustion (1)
9 any valid example, e.g. Spencer Silver (glue for sticky notes)/Ruth Benerito (wrinkle free cotton) (2)

(Total 20)

8 Earth and space

What I already know (page 52)

- The Sun is luminous because it gives out light. (1)
- There are eight planets in the Solar System. (1)
- Earth is the only planet known to support life. (1)
- The Moon orbits the Earth. (1)
- The Moon is not luminous, but it reflects light from the Sun. (1)
- Astronauts have travelled from the Earth to the Moon. (1)
- We should never look directly at the Sun even when wearing sunglasses. (1)

Planet facts (page 53)

1 Jupiter (1)
2 62 (1)
3 Uranus and Neptune (2)
4 2873 million km (1)
5 Venus (1)
6 Jupiter (1)
7 (a) the time taken for the planet to orbit the Sun (1)
  (b) Mercury is nearer the Sun, so its orbit is smaller/it has a shorter distance to travel. (2)
8 Neptune, Uranus, Saturn, Jupiter (2)
9 He is partly correct. The pattern is as he suggests, however Venus is much hoter than Mercury and this does not fit the pattern. (2)

Earth and Moon: day and night (page 54)

1 The rays should be drawn as straight lines with a ruler. (1)
2 The arrows should point away from the Sun, towards Earth. (1)
3 This should read, 'The Earth spins on its axis …' (1)
4 The shadow should be coloured so that the dark/light line is vertical, not along the axis. (1)
5 day (1)
6 night (1)

Sundial (page 55)

1 08:00 – 90 cm
  12:00 – 22 cm (2)
  (Allow any reasonable estimate.)
2 shadows correctly drawn on diagram (10:00 – 45 cm, 14:00 – 40 cm) (2)
3 Allow 75 – 85 cm. Shadow length shown on diagram should match the estimated length as written into the table. (2)

Research project: Moon diary (page 56)

The diary entry could include information such as the dusty surface scattered with rocks, no wind, no sound, dark sky, footprints, effect of lower gravity. The writing should show evidence of research having developed broader knowledge, and sources should be recorded in the bibliography.

Quick quiz: Earth and space (page 57)

1 sphere/spherical (1)
2 The ship would disappear gradually over the horizon. If the Earth was flat you would continue to see it get smaller but still visible. (2)
3 The Earth spins on its axis, so the Sun appears to move across the sky. (2)
4 At midday a shadow will appear short and roughly beneath/slightly to the north of the object. At late evening it will appear long and to the east of the object. (2)
5 As the Earth spins on its axis each part of the Earth moves from darkness to light. (2)
6 a year (1)
7 28 days (1)
8 the Sun and eight orbiting planets (2)
9 Milky Way (1)
10 gravity/gravitational force (1)
11 giving off light (1)
12 They reflect light from the Sun. (1)

(Total 17)

9 Forces

What I already know (page 59)

These two magnets will **attract** each other because the north-seeking pole of one is facing the **south**-seeking pole of the other. Opposite **poles** will attract one another. (3)

These two magnets will **repel** each other because the two **north**-seeking poles are facing each other. (2)

Magnets attract objects made from **magnetic** materials such as **iron** or steel. Repelling forces only occur between two **magnets**. (3)

In a tug-of-war, these two men are making **pulling** forces on the rope. They need plenty of **friction** between their shoes and the ground to stop them from slipping. Friction is caused when two **surfaces** rub together. It is greatest when the surfaces are **rough**. (4)

Sailing boats (page 60)

1 Identical boats, same material used for sails, hair dryer in the same position and on the same setting, start each boat at the same position, measure between the same start/finish points (i.e. front or back of boat). (6)
2 sail shape/size (1)
3 distance travelled by the boat (1)
4 Accept any sensible, clearly explained prediction (e.g. shape A, because it has the greatest surface area). (2)

Levers, pulleys and gears (page 61)

1 The long handle acts as a lever. A smaller effort is needed to turn on the tap because it is applied at the end of the lever. (2)
2 The arrow should show the smaller wheel turning anticlockwise. (1)
3 more than one turn (1)
4 The arrows should show a downward force on the bucket and a diagonally downward force on the rope. (2)
5 The force on the bucket should be labelled 'load' and the force on the rope should be labelled 'effort'. (2)
6 It is easier to pull downwards than to lift the load upwards. (1)

Forces: vocabulary (page 62)

1 force (1)
2 friction (1)
3 resistance (1)
4 streamlined (1)
5 upthrust (1)
6 fulcrum (1)
7 gears (1)
8 pulley (1)
9 reaction (1)
10 Highway (1)

Research topic: Samuel Plimsoll and safety at sea (page 63)

The mark is the Plimsoll line, also known as the International Load Line. It shows the maximum safe level to which a ship should be loaded. The heavier the ship, the lower it sits in the water. The Plimsoll line allows inspectors to know quickly if the ship is overloaded by looking at where the water level is in relation to the mark. The different lines show where the water level should be in different conditions, e.g. fresh/salt water, summer/winter, etc.

Answers might also include the work of Samuel Plimsoll in persuading Parliament to introduce a Royal Commission into shipping safety and the resulting Merchant Shipping Act.

Quick quiz: forces (page 64)

1 newton (1)
2 force meter/newton meter (1)
3 support/ reaction force (1)
4 frictional, gravitational, magnetic, upthrust (4)
5 using an arrow pointing in the direction that the force is acting (2)
6 levers, pulleys and gears (3)
   (Allow valid alternatives, e.g. inclined plane.)
7 slows them down (1)
8 to prevent skidding (1)
9 shaped such that air resistance/drag is minimised (1)
10 The faster the vehicle the longer the stopping distance/the slower the vehicle the shorter the stopping distance. (2)

(Total 17)

Quick quiz: properties of materials

Write *short* answers to the following questions to check how much you remember about properties of materials. You do not need to write full sentences.

1 What is meant by the term 'material'?

_____ (1)

2 Why is glass usually chosen as a material for windows?

_____

_____ (3)

3 Name one property that might need to be considered when choosing a material to make each of the following items. Choose a different property for each.

(a) a wall _____ (1)

(b) a cup _____ (1)

(c) a hosepipe _____ (1)

4 What are the two major groups into which materials are sorted?

_____ (2)

5 What word describes a material that is attracted to a magnet?

_____ (1)

6 Which is the only common magnetic metal?

_____ (1)

7 Explain why items such as paper clips, which are made of steel, are attracted to a magnet?

_____ (1)

8 How would we describe a material such as plastic, which is not attracted to a magnet?

_____ (1)

9 Describe briefly how you could compare the strength of two paper bags.

_____

_____

_____

_____

_____ (5)

(Total 18)

Check list: properties of materials	☺	☺	☹
I know	☺	☺	☹
that materials are what something is made from			
that the choice of material depends on its properties and the function of the object			
that properties of materials include hardness, strength, flexibility and magnetic behaviour			
that materials may be metals or non-metals			
that magnetic materials are attracted to a magnet			
that iron is the only common magnetic metal			
that steel is magnetic because it contains iron			
that materials that are not attracted to a magnet are described as non-magnetic			
I can	☺	☺	☹
design simple experiments to test the properties of materials			

5 Reversible changes

Look carefully at the pictures below. Each one shows a change of state occurring.

Under each picture, give the state (solid, liquid or gas) of the item shown.

On each arrow write the name of the change of state.

| condensation | evaporation | freezing | melting | solid | liquid | gas |

State: _____

State: _____

State: _____

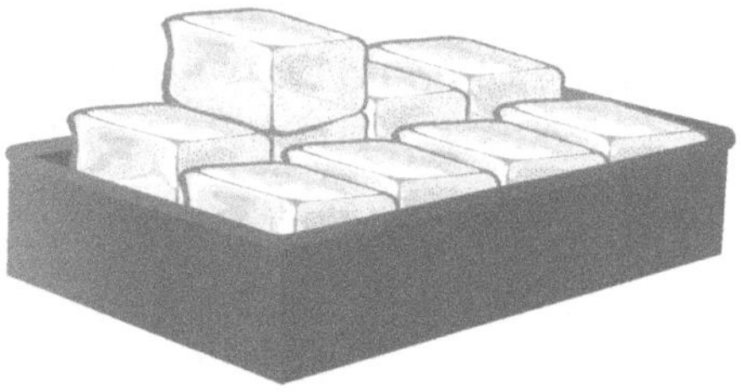

State: _____

State: _____

State: _____

State: _____

State: _____

(12)

Changes of state: the water cycle

Developing vocabulary

During the water cycle there are various changes of state.

Below are five statements describing changes that take place during the water cycle. Complete each one by filling in the gaps using words from the box.

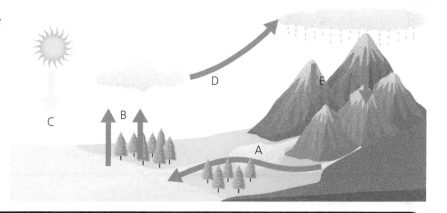

condenses	expands	vapour	evaporates	rivers
cracking	precipitation	cool	freeze	water

☐ The heat from the Sun warms the _____ in the sea. Some of the water _____ and rises up into the sky as water vapour.

☐ As the water _____ rises, it cools and _____ to form tiny droplets of liquid water. These come together to form clouds.

☐ Winds blow the clouds over the land. When they meet mountains, they rise and _____ even more. This makes the water droplets bigger, and they fall as rain, sleet or snow. This is called _____.

☐ Most of the water flows into _____, which carry it back to the sea. Some water soaks into the soil to become ground water.

☐ Some water may get trapped in crevices in the rocks on the mountain. In winter this water may _____. Water _____ when it freezes and this will push the crevices apart, _____ the rocks into smaller pieces.

(10)

Each statement above refers to one of the labelled stages in the picture. Place the correct letter from the diagram in the box by each statement.

(5)

Skill check

I know and understand the vocabulary for this topic. ☐

Garden hose disaster

Skill Using knowledge and understanding to explain an observation

Imagine you are the child in the pictures above. Write a letter to your grandmother, or some other person, describing the funny thing that happened when Dad tried to wash the car. Include a clear, scientific explanation of what happened.

_____

_____

_____

_____

_____

_____

_____

_____

_____

_____ (10)

Skill check

I have described and explained the event shown in the pictures, using appropriate scientific vocabulary. ☐

Dissolving experiment

Evaluating an experimental method and suggesting improvements

Organising data into a table

Two groups of children are investigating the rate at which salt crystals dissolve in water.

Each group describes their method.

Group A

We used small salt crystals. We put some crystals in some water and stirred until they had all dissolved. We used a timer to measure the time taken for them to dissolve. We did the experiment three times to check our results.

Our results: test 1 = 0 mins 35 seconds, test 2 = 0 mins 26 seconds, test 3 = 0 mins 58 seconds.

Group B

We used big salt crystals. We put 10 g of crystals in a beaker and added 100 cm^3 of water at 60 °C. We stirred until all the crystals had dissolved and timed how long it took. We did the experiment three times to check our results.

Our results: test 1 = 1 minute and 45 seconds, test 2 = 1 minute and 42 seconds, test 3 = 1 minute and 47 seconds.

1 Fill in the results from these two experiments in the table below.

	Time taken to dissolve, in minutes and seconds		
	Test 1	Test 2	Test 3
Group A			
Group B			

(4)

2 (a) Which group's data was unreliable? _____ (1)

(b) Suggest two ways in which they could improve their method to make their results more reliable.

_____

_____ (2)

3 Suggest one safety precaution that the children should take when doing this experiment.

_____ (1)

4 Group B weighed the solution at the end of each test. If 1 cm^3 of water weighs 1 g, what would be the final mass of their solution? Show your working.

_____ (2)

Skills check

I can spot errors in an experimental method and suggest improvements. ☐

I can organise data into a table. ☐

Solutions: crossword

Skill Developing vocabulary

Across

1 A _____ is a mixture of a liquid and a soluble solid.

3 When no more solute can dissolve, the solution is said to be _____

4 Sugar and salt dissolve in water because they are _____

7 Tiny pieces of an insoluble solid floating (suspended) in a liquid are called a _____

8 In a solution, the liquid part is the _____

Down

2 Chalk is _____ so it is not able to dissolve in water.

5 When salt is mixed with water it seems to disappear because it has _____

6 A _____ is the solid part of a solution.

(8)

Skill check

I know and understand the vocabulary for this topic. ⬜

Quick quiz: reversible changes

Write *short* answers to the following questions to check how much you remember about reversible changes. You do not need to write full sentences.

1 Explain why the freezing of water is a reversible change.

_____ (1)

2 What name is given to the following changes of state:

(a) solid to liquid _____ (1)

(b) gas/vapour to liquid _____ (1)

3 What is happening when a puddle disappears from the playground?

_____ (1)

4 What name do we give to the change that occurs in water at 100 °C?

_____ (1)

5 What is the freezing point of water? _____ (1)

6 What property of water as it freezes, causes pipes and rocks to crack?

_____ (1)

7 What does the term 'insoluble' mean?

_____ (1)

8 What is formed when salt dissolves in water?

_____ (1)

9 (a) What name do we give to the solid part of a solution?

_____ (1)

(b) What name do we give to the liquid part of a solution?

_____ (1)

10 Which would you expect to dissolve more quickly, large salt crystals or small salt crystals?

_____ (1)

11 How does the temperature of the solvent affect the rate of dissolving?

_____

_____ (2)

12 What effect does stirring have of the rate of dissolving?

_____ (1)

13 What will be the mass of a solution made from 100 cm³ (100 g) of water and 20 g of salt?

_____ (1)

(Total 16)

Check list: reversible changes

I know	☺	☺	☹
that dissolving, melting, boiling, condensing, freezing and evaporating are all reversible changes			
that heating and cooling cause changes of state			41
that water expands on freezing, and that this can cause pipes to burst and rocks to crack			
that soluble solids dissolve in liquids to form solutions			
that insoluble solids are not able to dissolve			
that a solution is formed from a solvent (liquid) and a solute (soluble solid)			
that temperature of the solvent, particle size of the solute and stirring are all factors that affect the rate of dissolving			
that the mass of a solution is equal to the mass of the solvent plus the mass of the solute			
I can	☺	☺	☹
describe how to carry out a fair test to compare rates of dissolving			

6 Separating mixtures

What I already know

The boxes below each contain half a sentence. Draw lines to join the pairs to make full sentences.

The particles in a soil sample can be separated	magnetic material.
Iron is a	muesli
A mixture of steel and plastic paper clips could be separated	it evaporates.
An example of a mixture is	but stones will sink.
Air is	using sieves.
When water is heated	a mixture of gases.
A solution contains	using a magnet.
Leaves and sticks will float on water	a liquid and a dissolved solid.

(8)

Mixtures are all around us

Skill   Sorting and classifying

1   The words in the box include a number of different materials.

Some are mixtures and some are not.

Write each one into the correct column in the table below.

| air | coffee (drink) | muesli | oxygen gas | sea water |
| chocolate | iron | orange squash | salt | soil |

Mixtures	Pure materials

(10)

2   The mixtures shown below each have an odd one out. For each mixture, select the odd one out and then say why you think it does not fit with the others.

Mixture	Odd one out	Why
Flour, sand, eggs, butter		
Water, salt, seaweed, sugar		
Gravel, leaves, silt, glass		

(3)

Skill check

I can identify common mixtures. ☐

I can identify and explain things that do not fit in a group. ☐

Separating mixtures

Some children have made a saturated salt solution by adding salt to water until no more can dissolve. Their teacher challenges them to get the salt back from the solution. They must achieve this before the end of the lesson.

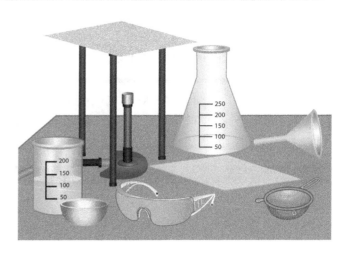

The teacher gives them the apparatus shown. They can choose which pieces to use.

Describe clearly how they should do this. Remember to include any special instructions to make sure that they carry out the experiment safely.

_____

_____

_____

_____

_____

_____

_____

_____

_____

_____

_____

_____

Skill check

I can describe an experimental method clearly.

I can identify important actions to make sure that the experiment is safe.

Drawing science diagrams

Skill   Drawing and labelling neat science diagrams

In this exercise you are going to practise the important skill of drawing and labelling really neat science diagrams. Remember to use a sharp pencil and to draw all straight lines using a ruler.

1   Draw a science diagram of each of the following pieces of apparatus.

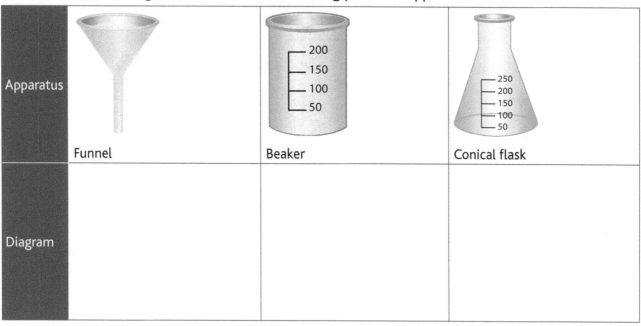

Apparatus	Funnel	Beaker	Conical flask
Diagram			

(6)

2   In the space below draw a neat, labelled diagram to show the apparatus used to filter a mixture of sand and water. Label the filtrate and residue.   (5)

Skill check

I can draw neat science diagrams. ☐

I can label my diagrams neatly. ☐

Quick quiz: separating mixtures

Write *short* answers to the following questions to check how much you remember about separating mixtures. You do not need to write full sentences.

1 A mixture of a solid and a liquid can be separated by two different methods, decanting and filtering. Which would be the best to choose for separating the following mixtures?

(a) A suspension of chalk in water _____ (1)

(b) A mixture of gravel and water _____ (1)

2 After filtering some muddy water, what name would we give to:

(a) the mud remaining in the filter paper? _____ (1)

(b) the clean water in the flask or beaker? _____ (1)

3 (a) Which method would you use to recover salt crystals from a salt solution? (Underline the correct answer.)

decanting   evaporating   filtering   sieving   (1)

(b) Explain briefly how the chosen method works.

_____

_____ (2)

4 Explain briefly why a salt solution should never be heated until it is completely dry.

_____ (1)

(Total 8)

Check list: separating mixtures			
I know	☺	😐	☹
that filtering and decanting are methods used to separate a solid from a liquid			
the meaning of the terms 'filtrate' and 'residue'			
how to recover a dissolved solid from a solution by evaporation			
that salt solutions should not be heated until completely dry for safety reasons			
I can	☺	😐	☹
use my knowledge of solids, liquids and gases to suggest methods for separating mixtures			
select appropriate apparatus for separation activities			
draw and label neat science diagrams			

7 Chemical changes

What I already know

Changes to materials are occurring around us all the time. Some of them are reversible – the material can be changed back into its original state. Some are non-reversible – the material cannot be changed back into its original state.

The pictures below show some changes taking place. Use your existing knowledge to identify which of the changes are reversible and which are non-reversible. Draw a straight line from each change to the correct word.

firework exploding

REVERSIBLE

sugar dissolving

snowman melting

leaves rotting

egg frying

NON-REVERSIBLE

wood burning

hair drying

trees growing

(8)

An important chemical change: combustion

Skill Choosing correct measuring apparatus
Identifying safety procedures
Drawing conclusions from data in a bar chart

Some scientists are investigating some different fuels.

They put a small amount of a fuel into a metal container.

They place the container under a beaker of water.

They set light to the fuel and measure the change in temperature of the water in the beaker.

They do the same for each fuel. To make it a fair test they make sure that they use the same mass of fuel and the same volume of water each time.

1 What piece of apparatus should they use to measure

   (a) the mass of the fuel _____ (1)

   (b) the volume of the water _____ (1)

   (c) the temperature of the water _____ (1)

2 The scientists carry out their investigation safely. Suggest one thing they would do to protect themselves from harm.

   _____ (1)

The bar chart shows the results from their experiment.

3 Something very important is missing from the bar chart. What is this?

   _____ (1)

4 The statements below show some conclusions that might be drawn from the experiment. Decide whether each one is **true**, **false**, or if you **cannot tell** from the bar chart above and place a tick in the correct column.

	True	False	Cannot tell
Fuel A heated the water most			
The temperature rise in the water for Fuel B was 20°C			
Fuel C heated the water less than Fuel B			
All the fuels heated the water by more than 10°C			

(4)

Skill check

I can select the correct apparatus for measuring quantities. ☐

I can describe how to stay safe when carrying out combustion experiments. ☐

I understand what conclusions can be drawn from a bar chart. ☐

Another chemical change: rusting

Skill Using existing knowledge and understanding to predict the outcome of an experiment.

The children have each been given an iron nail. Their teacher challenges them to do something to their nail to stop it from rusting. All the nails will be left in the playground for a week. The weather is expected to be rainy.

1 Each child has chosen a different method. Look carefully at their ideas. In the box by each child, predict how well their method would prevent the nail from rusting. Remember to explain your ideas.

Idea	Prediction
I will wrap my nail in sticky tape.	
I will paint my nail with glue.	
I will dip my nail in some cooking oil.	
I will cover my nail with finger nail varnish.	

(8)

2 Their teacher adds a nail without any protection so that they can compare it with their experiments. What would this extra nail be called? _____ (1)

Skill check

I can use my existing knowledge to predict the outcome of an experiment. ☐

Quick quiz: chemical changes

Write *short* answers to the following questions to check how much you remember about chemical changes. You do not need to write full sentences.

1 Chemical changes are non-reversible. In what other way do they differ from reversible changes such as melting and evaporation?

_____ (1)

2 Are the following changes naturally occurring or man-made (synthetic)?

(a) making concrete _____ (1)

(b) cooking food _____ (1)

(c) ripening fruit _____ (1)

(d) rusting _____ (1)

3 What two things are necessary to cause rusting of iron?

_____ (2)

4 Give three methods that might be used to prevent rusting.

_____ (3)

5 What is a fuel?

_____

_____ (2)

6 What is a fossil fuel?

_____

_____ (2)

7 Give an example of a fossil fuel that is

(a) a solid _____ (1)

(b) a liquid _____ (1)

(c) a gas _____ (1)

8 What is the term is usually used by scientists to mean burning?

_____ (1)

9 Give an example of a new material that was created by experimental work by a scientist.

_____ (2)

(Total 20)

Check list: chemical changes

I know	☺	😐	☹
that non-reversible (chemical) changes result in the formation of new materials that may be useful			
that making concrete, cooking food, ripening fruit, setting superglue, rusting and combustion are examples of everyday non-reversible (chemical) changes			
that chemical changes may be the result of naturally-occurring or man-made (synthetic) processes			
that chemists use chemical changes to create new materials			
that air and water are both needed for rusting to occur			
that oiling, painting, galvanising and covering with plastic can prevent rusting			
that a fuel is a material that is burnt to release useful energy			
that fossil fuels were formed millions of years ago from the remains of plants and animals and are not renewable			
that coal (solid), oil (liquid) and natural gas are fossil fuels			
that combustion (burning) is a non-reversible reaction that results in the formation of new materials (carbon dioxide and water vapour)			
I can	☺	😐	☹
give an example of how a chemist created a new useful material			

What I already know

The boxes below each contain half a sentence about Earth and space.

Draw straight lines to join pairs of boxes to make seven complete sentences.

The Sun is luminous

known to support life.

There are eight planets

in the Solar System.

because it gives out light.

Earth is the only planet

Astronauts have travelled

The Moon orbits

it reflects light from the Sun.

The Moon is not luminous, but

from the Earth to the Moon.

the Earth.

We should never look directly at the Sun

even when wearing sunglasses.

(7)

Planet facts

Skill Reading and understanding information given in a table

The table shows some facts about the planets of our solar system. Use information from the table to help you to answer the questions below.

Planet	Diameter, in km	Number of known moons	Year length, in Earth days	Day length, in hours	Average surface temperature, in °C	Distance from Sun, in millions of km
Mercury	4879	0	88	4223	167	58
Venus	12104	0	225	2802	464	108
Earth	12756	1	365	24	15	150
Mars	6792	2	687	25	−65	228
Jupiter	142984	67	4331	10	−110	779
Saturn	120536	62	10747	11	−140	1434
Uranus	51118	27	30589	17	−195	2873
Neptune	49528	15	59800	16	−200	4495

1 Which is the largest of the planets? _____ (1)

2 How many moons are known to orbit Saturn? _____ (1)

3 Which planets have a lower average temperature than Saturn?

_____ (2)

4 How far from the Sun is Uranus? _____ (1)

5 Which planet is closest in size (diameter) to Earth? _____ (1)

6 Which planet has the shortest day? _____ (1)

7 (a) What do we mean by the term 'year'?

_____ (1)

(b) Suggest a reason why Mercury's year is so much shorter than Earth's year.

_____ (2)

8 List the planets larger than Earth, in order of size, smallest first.

_____ (2)

9 Alex says, 'The nearer to the Sun, the warmer the planet.' Is he correct? Explain your answer.

_____

_____ (2)

Skill check

I can find and interpret information from a table to answer questions. ☐

Earth and Moon: day and night

A student was asked to draw a diagram to show how day and night occur. His diagram is shown below.

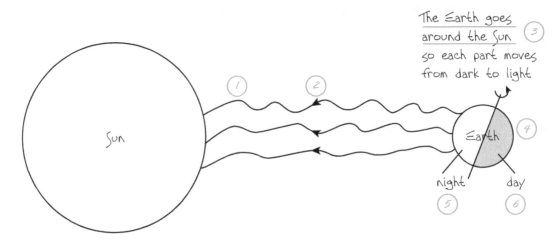

The teacher has marked his work by showing six places where the diagram needs improving.

Suggest what the student should do to improve his work at each of the marked places.

1 _____ (1)

2 _____ (1)

3 _____ (1)

4 _____ (1)

5 _____ (1)

6 _____ (1)

Skill check

I correctly identified the error at each marked position. ☐

I suggested ways to improve the diagram. ☐

Sundial

Interpreting data in an unusual format

A school playground contains a simple sundial. The gnomon is a vertical stick. The times are marked on the ground as shown in the diagram below. The children use chalk to record the length of the shadow on the playground surface each hour on a sunny day. They record the length of the shadow in a table.

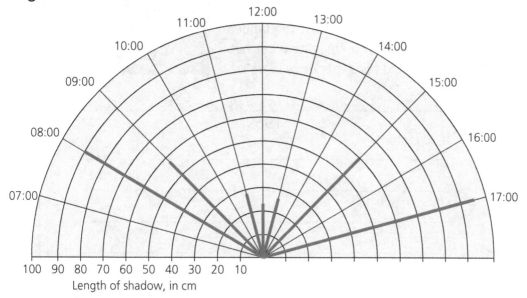

Length of shadow, in cm

Time	Length of shadow, in cm
08:00	
09:00	58
10:00	45
11:00	28
12:00	
13:00	27
14:00	40
15:00	60
16:00	did not measure
17:00	95

1 The children have left out some of their results. Look at the diagram of the sundial to estimate the length of the shadow at 08:00 and 12:00 and complete the table. (2)
2 Use data from the table to draw in the correct length of shadow at 10:00 and 14:00. (2)
3 The children were unable to measure their shadow at 16:00. Estimate how long the shadow would have been. Write your estimate in the table and draw the shadow on the diagram. (2)

Skill check

I have understood how the data was displayed and am able to use the format to interpret the data and make predictions. ⬜

Research project: Moon diary

Skill Using books and the internet to carry out research

Recording sources of information in a bibliography

The first person to step onto the surface of the Moon was Neil Armstrong in 1969. The Moon's surface is very different to the surface of the Earth. Use books and the internet to find out about what he might have seen and felt when he left the lander and stepped onto the Moon for the first time. Imagine what he might have written in his diary to record his experience.

Monday   1969

21

July

Bibliography

_____

_____

Skill check

I have found suitable sources of information for my research. ☐

I have recorded my sources of information. ☐

8 Earth and space

56

Quick quiz: Earth and space

Write *short* answers to the following questions to check how much you remember about Earth and space. You do not need to write full sentences.

1 What shape are the Sun, Earth and Moon?

_____ (1)

2 Describe how watching a sailing ship sail away from port might show observers that the Earth is not flat.

_____

_____ (2)

3 Why does the position of the Sun seem to change during the day?

_____ (2)

4 What two changes might be observed in the shadow of an object as seen at midday and late evening?

_____

_____ (2)

5 Why do we experience day and night?

_____ (2)

6 What do we call the time taken for the Earth to orbit the Sun once?

_____ (1)

7 How long does the Moon take to orbit the Earth?

_____ (1)

8 What is the solar system?

_____ (2)

9 What is the name of our galaxy?

_____ (1)

10 Which force pulls objects towards Earth and keeps the planets in orbit?

_____ (1)

11 Stars, including the Sun, are luminous. What does this mean?

_____ (1)

12 How do we see space objects such as the Moon and planets?

_____ (1)

(Total 17)

Check list: Earth and space

I know	☺	☺	☹
that the Sun, Earth and Moon are approximately spherical			
how evidence from ships going over the horizon and orbiting spacecraft confirm that the Earth is spherical			
how the position of the Sun appears to change during the day and how shadows change as this happens			
how day and night are caused by the Earth spinning on its own axis			
that the Earth orbits the Sun once each year			
that the Moon is a natural satellite of the Earth and takes approximately 28 days to orbit the Earth			
that the planets of the solar system orbit the Sun			
that the solar system is part of the Milky Way galaxy			
that the Universe contains a huge number of galaxies and that the distances between bodies in space are enormous			
that objects are pulled downwards because of the gravitational attraction between them and the Earth			
that it is gravitational forces which keep the Moon in orbit around the Earth and planets in orbit around the Sun			
that the Sun and other stars are light sources and that planets and other bodies are seen by reflected light			
I can	☺	☺	☹
describe how the work of scientists such as Ptolomy, Alhazen and Copernicus helped to develop our understanding of the Universe			
describe how the work of scientists such as Isaac Newton and Galileo Galilei helped develop the theory of gravitation			

9 Forces

What I already know

Use words from the box to complete the sentences below.

attract	friction	iron	magnetic
magnets	north	poles	pulling
repel	rough	south	surfaces

These two magnets will _____ each other because the north-seeking

pole of one is facing the _____-seeking pole of the other. Opposite

_____ will attract one another. (3)

These two magnets will _____ each other because the two

_____-seeking poles are facing each other. (2)

Magnets attract objects made from _____ materials such

as _____ or steel. Repelling forces only occur between two

_____. (3)

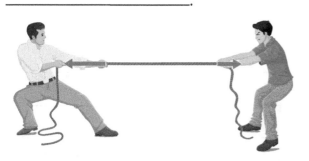

In a tug-of-war, these two men are making _____ forces on the rope. They

need plenty of _____ between their shoes and the ground to stop them

from slipping. Friction is caused when two _____ rub together. It is greatest

when the surfaces are _____. (4)

Sailing boats

Some children are investigating air resistance. They each have a small boat with a mast.

Each child makes a different shape of sail for their boat.

They have a long piece of gutter pipe full of water. They can use a hair dryer to blow the boats along to see which one travels furthest from the hair dryer.

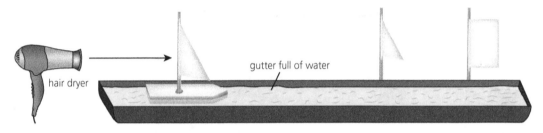

hair dryer

gutter full of water

1 To find out which sail shape is best, they need to carry out a fair test. What variables need to be controlled to make their test fair?

_____

_____

_____

_____ (6)

2 What is the independent variable in this experiment? _____ (1)

3 What is the dependent variable in this experiment? _____ (1)

Three of the sails are shown on the right.

4 Predict which sail shape would help the boat travel furthest. Explain your answer.

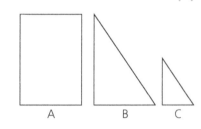

A   B   C

_____

_____

_____ (2)

Skill check

I can identify the different types of variable, including the control variables to make a fair test.	⬜
I can use my scientific knowledge and understanding to make and explain a sensible prediction.	⬜

Levers, pulleys and gears

Skill Understanding simple mechanisms

1 The picture shows two types of tap. Explain how the design of the right-hand tap makes it easier for someone to turn the tap on or off.

_____

_____

_____ (2)

2 The picture shows two gear wheels. The larger wheel is turned clockwise.

Draw an arrow on the diagram to show the direction of movement of the smaller wheel. (1)

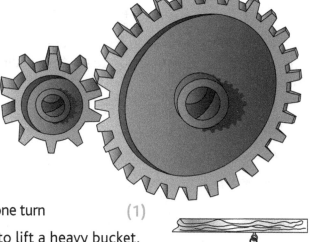

3 Underline the correct phrase to complete the following sentence.

If the larger wheel is turned through one complete turn, the smaller wheel will turn …

less than one turn   one turn   more than one turn   (1)

4 The picture shows a pulley being used to help to lift a heavy bucket.

(a) Draw arrows on the diagram to show the direction of the force needed to lift the bucket and the gravitational force of the bucket. (2)

(b) Label the arrows using the terms 'effort' and 'load'. (2)

(c) Explain briefly how this mechanism helps the person to lift the heavy load.

pulley

_____

_____ (1)

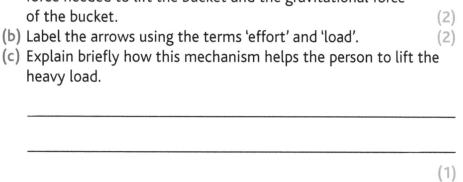

Skill check

I can explain how simple mechanisms work. ⬜

Forces: vocabulary

```
V  D  R  X  Q  V  D  S  E  A  P  J
S  T  R  E  A  M  L  I  N  E  D  B
P  H  I  E  S  Z  P  E  R  W  U  J
L  G  F  H  U  I  I  K  H  O  Z  W
E  E  R  I  P  V  S  T  R  G  Z  D
F  A  I  G  T  X  V  T  E  J  X  S
U  R  C  H  H  P  I  Z  A  G  Y  B
L  S  T  W  R  U  Z  E  C  N  G  F
C  A  I  A  U  L  R  P  T  L  C  D
R  M  O  Y  S  L  O  P  I  J  O  E
U  U  N  V  T  E  V  F  O  R  C  E
M  E  C  M  P  Y  W  R  N  V  S  Z
```

Find words in the wordsearch to complete the definitions. Highlight each word neatly when you find it and make sure that you check the spelling when writing it.

1 A _____ is a pushing, pulling or twisting action on an object.   (1)

2 The force of _____ is caused when surfaces rub together.   (1)

3 As objects move through the air they are slowed down by air _____   (1)

4 Objects that are shaped so that air or water passes over them smoothly are said to be

   _____   (1)

5 The upward force of water that can make an object float is called _____ (1)

6 The fixed point about which a lever acts is called the _____   (1)

7 Toothed wheels that are used to make cycling more efficient are called _____ (1)

8 A heavy load can be lifted with a downward force on a rope that passes over a

   _____   (1)

9 The opposing force felt when an object is pushed, pulled or twisted is called the

   _____ (or support) force.   (1)

10 The effect of speed on the stopping distance of a car is given in the

   _____ Code.   (1)

Skill check

I know and understand the vocabulary for this topic. ☐

Research project: Samuel Plimsoll and safety at sea

Skill Using books and the internet to carry out research

Recording sources of information in a bibliography

In the 19th Century there were many cargo ships at sea, carrying goods around the world. Some ship owners were greedy, and often ships were overloaded and unstable. Working on a merchant vessel was a dangerous job for seamen, and many lives were lost at sea.

In 1874 a law was passed, introducing this special mark that must be painted on the hull of all ships. It is now used on ships from all round the world.

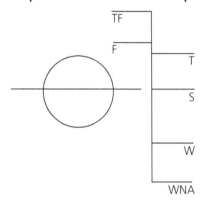

Find out what this mark is called and how it helps to make ships safer at sea.

_____

_____

_____

_____

_____

_____

_____

_____

_____

Bibliography

_____

_____

Skill check

I have found suitable sources of information for my research. ☐

I have recorded my sources of information. ☐

Quick quiz: forces

Write *short* answers to the following questions to check how much you remember about forces. You do not need to write full sentences.

1 What is the unit of force?

_____ (1)

2 What apparatus is used to measure forces?

_____ (1)

3 What do we call the opposing force that is felt when an object is pushed or pulled?

_____ (1)

4 Name four other types of force.

_____ (4)

5 How do we show the direction of a force in a diagram?

_____ (2)

6 Name three mechanisms that can be used to allow a small force to have a greater effect.

_____ (3)

7 What effect do frictional forces have on the movement of objects?

_____ (1)

8 Why is it important to have friction between tyres and the road surface?

_____ (1)

9 What is meant by the term 'streamlined'?

_____ (1)

10 What effect does the speed of a vehicle have on the stopping distance?

_____ (2)

(Total 17)

Check list: forces			
I know	☺	☺	☹
that the unit of force is the newton (N) and that forces can be measured using a force meter (newton meter)			
that when objects are pushed or pulled, an opposing pull or push can be felt, and that this is called a support (reaction) force			
that types of force also include frictional (including air resistance), magnetic, gravitational and upthrust			
how to use arrows to show the direction in which forces are acting			
that levers, pulleys and gears allow a smaller force to have a greater effect			
how frictional forces, including air resistance, affect movement			
that friction between tyres and the road slows the movement of a vehicle and helps to prevent skidding			
that streamlining reduces the drag caused by air resistance			
about stopping distances, as listed in the Highway Code			
I can	☺	☺	☹
measure forces using a force meter			